I Have a Watch!

written by
Deborah Williams

Illustrated by
Dennis Graves

KAEDEN BOOKS™

This is Kelly's watch.

What time does it say?

One o'clock.

This is Mark's watch.

What time does it say?

Two o'clock.

This is my teacher's watch.

What time does it say?

Three o'clock.

This is Dad's watch.

What time does it say?

Four o'clock.

This is my watch.

What time does it say?

Five o'clock.
It's time to eat!